HOW TO GROW GREAT ALFALFA

& OTHER FORAGES

Dr. Harold Willis

A Biological Farmer's Guide

HOW TO GROW GREAT ALFALFA

& OTHER FORAGES

Dr. Harold Willis

A Biological Farmer's Guide

Acres U.S.A.
Viroqua, Wisconsin U.S.A.

About Acres U.S.A.

Founded in 1971 by Charles Walters, Acres U.S.A. emerged from the need to promote ecological farming practices in a time when industrial agriculture was heavily reliant on synthetic fertilizers and pesticides. Inspired by figures like Rachel Carson and Dr. William Albrecht, Walters used the magazine, and later books and conferences, to advocate for sustainable agriculture that prioritized soil health and natural processes. Acres U.S.A. provided a platform for these ideas and helped to popularize alternative methods like cover cropping and integrated livestock management.

Though the agricultural landscape still relies heavily on conventional methods, Acres U.S.A. has been instrumental in the growing movement towards regenerative agriculture. By disseminating knowledge and supporting eco-conscious farmers, the company continues to champion sustainable practices through its publications, conferences, and online resources, contributing to a shift towards a more environmentally sound approach to farming.

Find Out More About Acres U.S.A.

Subscribe To the Online Magazine
(https://members.acresusa.com/)

Attend The Eco-Ag Event
(https://www.acresusa.com/events/)

Visit The Acres U.S.A. Bookstore
(https://bookstore.acresusa.com/)

Join The Free Newsletter
(https://mailchi.mp/acresusa/newsletters)

How to Grow Great Alfalfa
& Other Forages

Acres U.S.A.
PO Box 351
Viroqua, Wisconsin 54665 U.S.A.
1 (970) 392-4464
info@acresusa.com • www.acresusa.com

Printed in the United States of America

Publisher's Cataloging-in-Publication

Willis, Harold L., 1940-
How to grow great alfalfa . . . and other forages / Harold L. Willis. Viroqua, WI, ACRES U.S.A., 2008
iv, 48 pp., 23 cm.
Includes Index
Includes Bibliography
Includes Illustrations
ISBN 978-1-60173-003-9 (trade)

SB205.A4 W55 2008   633.31

ADDENDUM TO HOW TO GROW GREAT ALFALFA & OTHER FORAGES

This book was published in 1983, and some product names mentioned in the book have changed since then.

One product, PhytoMast®, is the officially registered name for what was called Phyto-Mast in this book. The product is an herbal blend of thymol (antiseptic) balanced with angelica, licorice, wintergreen, olive oil, and vitamins A, D, and E. It is packaged in aseptic infusion tubes (with alcohol pads) to be used as intramammary infusion for healthy cows at dry-off. Peer-reviewed, published articles in studies done with dairy cows and goats have shown its utility and safety as a dry-off product. PhytoMast® is labelled specifically with commercial Grade A inspected dairy farms in mind so the farm can meet the regulations of the Pasteurized Milk Ordinance (PMO) item 15r (Storage of Drugs and Chemicals).

Along the way, I enhanced the formula and added cinnamon bark oil (CBO) due to many published studies in the literature demonstrating the destructive effects of CBO upon biofilms that various pathogenic germs form to protect themselves. To keep these similar yet different products separate, the name UdderWell is used for the formula with CBO. UdderWell should not be on the shelf of Grade A inspected dairy farms. In other words, UdderWell is labelled for those who are raising animals for themselves and not for commercial use. The difference between the two products, other than the presence of CBO, is strictly a labeling issue and not a quality or concentration difference.

UdderWell, GetWell, BreatheWell, EatWell, LivWell, BreedWell, and FeelWell are available at www.reverencefarms.com

Pages	New product names
331	**UdderWell/ PhytoMast®** for step (3)
206, 220, 238, 243–245, 269, 290, 373	**GetWell** for herbal antibiotic tincture; for herbal antibiotic mix; for Phytobiotic; for tincture of garlic, echinacea, goldenseal, wild indigo, and barberry
210, 218, 221, 223, 225, 233, 373	**EatWell** for tincture of ginger, gentian, nux vomica, and fennel; for Digestive Ø; for Phyto-Gest
210, 214, 217, 373	**LivWell** for tincture of red root, celandine, milk thistle, dandelion, and oregon grape root; for Phytonic
373	**FeelWell** for Phyto-gesic; for herbal pain formula tincture
284, 289–292, 373	**BreedWell** tincture / **HeatSeek** powder for "Spectra 305"
232, 238, 239, 243–245, 302, 327	**AmpliMune®** for ImmunoBoost; see www.novavive.ca
232, 238, 244, 245	**BoviSera®** for Bo-Bac-2X/ Quatracon/ BoviSera; see https://colorado-serum-com.3dcartstores.com

CONTENTS

FOREWORD ix

CHAPTER 1: What's it all About? 1

CHAPTER 2: Care and Feeding 7

CHAPTER 3: Quality 15

CHAPTER 4: Harvesting 29

CHAPTER 5: Problems 33

CHAPTER 6: Varieties 43

REFERENCES: 45

INDEX: 46

FOREWORD

Agriculture has undergone considerable change in the last 30-40 years. Yields have increased. So have fertilizer and pesticide use, but crop quality has gone downhill. Average farm size has increased, while the number of farms has shrunk. Tractors are bigger, but the soil is harder. And many farmers are having trouble raising alfalfa.

Could something be wrong? Are you really being told the best way to farm?

Consider this. Farmers and ranchers are working with plants and animals—living creatures. All of the marvelous variety of plants and animals, birds and butterflies on the earth were created to function best under certain conditions and to obey certain laws, which are just as exacting as the laws of physics and chemistry. In fact, any living organism—or any cell from any living organism—is infinitely more complex than any man-made machine or computer. And just as man's creations require certain materials, conditions, and types of energy to work properly, so do the crops and animals which are your source of income.

Throw a wrench into a jet engine or hit a watch with a hammer and you have problems. *Is it any wonder that agriculture today is having problems when so many farmers are trying to raise crops and animals while violating some of the biological laws of healthy growth and nutrition?*

Sure, you can get crops and animals to survive and give you some production, and maybe even record-breaking yields . . . but *are they at peak health, and are the products they produce of top quality?*

The emphasis today is almost always on quantity, *not quality*. Both crops and animals are over-fertilized, gorged, activated, supplemented, stimulated, and prodded—to produce *more* . . . and MORE!

A growing number of farmers are getting off this dizzying not-so-merry-go-round. They are learning that they can grow excellent yields of high quality crops and raise healthy, productive animals.

If you would like to know how to raise abundant, nutritious forage crops which your animals can produce more on, while eating less—read on . . .

CHAPTER 1

What's it all About?

A lfalfa has been called the "queen of forages" because of its remarkable ability to produce high yields of nutritious, palatable forage under a wide range of soil and climatic conditions.[1]

Alfalfa and other forage crops are an important and vital part of the agriculture of the United States, especially in the high dairy areas of the Great Lakes region and the Northeast, as well as along the Pacific West Coast. Forages are also important wherever livestock are fattened. In 1969, the total acreage harvested for hay and seed in the U.S. was 27.1 million acres, of which 26.6 million were used for alfalfa. Out of this, over 60% was grown in the Great Lakes region.[2]

Kinds of forages. Besides alfalfa, other forage species most often grown include red clover, sweetclover, birdsfoot trefoil, Ladino clover, white clover (all of those are legumes), plus smooth bromegrass, timothy, bluegrass, reed canarygrass, and orchardgrass (the latter five are grasses, in a different plant family than the legumes).

Other forage crops that are not grown as commonly or that are restricted to certain parts of the country include, among the legumes: alsike clover, sour clover, crimson clover, lespedezas, vetches, soybeans, field peas, cowpeas, peanut vines, and kudzu; and among the grasses: fescues, redtop, meadow foxtail, Sudangrass, Johnsongrass, sorghum and its hybrids, millet, proso, tall oatgrass, wheatgrasses, bluestems, grama grasses, buffalograss, switchgrass, lovegrass, ryegrasses, needle grasses, Bahiagrass, Bermudagrass, carpetgrass, Dallisgrass, oats, barley, wheat, rye, and corn (maize).

Since alfalfa is by far the most widely grown forage species, this book will mainly deal with alfalfa, although most of the legume forages

---

[1] J.C. Burton, p. 229 in *Alfalfa Science and Technology*, 1972.
[2] J.L. Bolton, B.P. Goplen, & H. Baenziger, p. 24 in *Alfalfa Science and Technology*, 1972.

are about the same in their growth requirements. To simplify matters, we will briefly list the characteristics of the non-alfalfa forages first and then concentrate on alfalfa.

Legumes

The plants in the legume family have the distinct ability to provide a home for a type of nitrogen-fixing bacteria, *Rhizobium*, in the swollen nodules that can form on the legume's roots. These bacteria live in a symbiotic relationship with the legume and are able to capture (fix) nitrogen gas from the air and change it into ammonia, which the legume uses to produce proteins. There are many varieties or strains of the bacteria, and only certain ones can successfully form nodules on a certain species or variety of legume. Although the bacteria are generally common in fertile soil (which has not been sterilized by toxic chemicals), it is best to plant seed that has been inoculated with the right strain of bacteria to insure successful nodule formation.

Most legume forages are perennial plants (they live more than one year), although some sweetclover and alfalfa varieties are annual (live one year). Most legumes are characterized by deep taproots and growth of several stems from a crown region near ground level. The stems grow

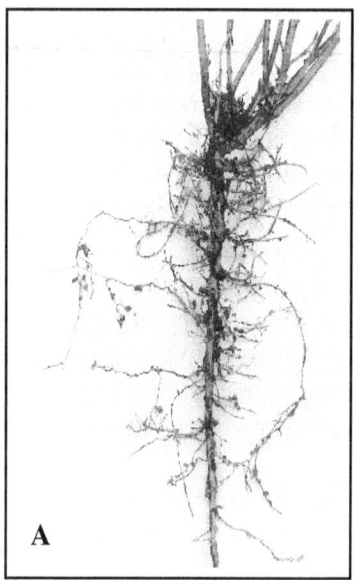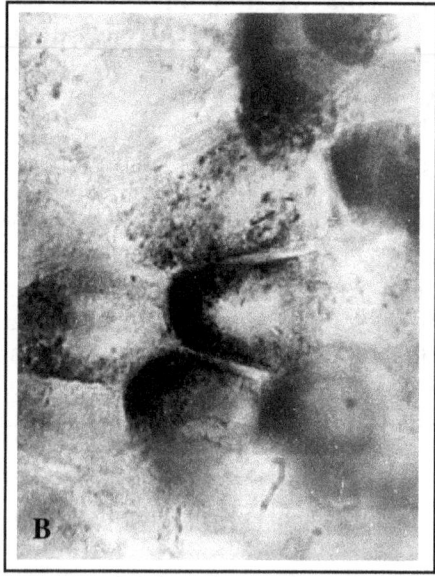

A. Root nodules of alfalfa; B. nitrogen-fixing bacteria inside alfalfa root nodule cells, magnified 100 times under light microscope.

and elongate at the tips, but when the crop is harvested or grazed, new stems grow from buds in the crown.

Birdsfoot trefoil. Long-lived perennial; moderate-yielding with good midseason growth and late maturity; fair drought-tolerance and generally poor winter hardiness; does well on poor soils; tolerates continuous but not close grazing; difficult to establish. A related species is called big trefoil.

Red clover. Short-lived perennial (2 years in North) or annual (South); moderately high yielding with fair midseason growth; fair drought-tolerance and winter hardiness.

Red clover

Sweetclover. Biennial (lives 2 years) or annual; high-yielding with only moderate top growth the first season and little late growth the second season in 2-year varieties; good drought-tolerance and winter hardiness; does well on poor soils; not very palatable to livestock because of coumarin content; makes poor hay; does not tolerate close cutting or grazing.

White clover and *Ladino clover.* Rapid-growing perennial (Ladino is short-lived); low-yielding with early spring and poor midseason growth; poor drought-tolerance and good to moderate (Ladino) winter hardiness; tolerates continuous grazing.

Grasses

Grasses are characterized by comparatively shallow, diffuse root systems (with many roots, but no main taproots). The growing points for leaves and seed stalks are at ground level, so cutting or grazing will not injure them. Most forage grasses are perennial, while many weedy grasses are annual, as are crop grains (oats, wheat, corn, etc.).

Bluegrass, Kentucky bluegrass. Long-lived cool season perennial; low-yielding with early spring and poor midseason growth; poor drought-tolerance and very good winter hardiness; tolerates continuous grazing.

Bromegrass, smooth bromegrass. Long-lived cool season perennial; high-yielding with moderate spring and fair midseason growth; moderately good drought-tolerance and very good winter hardiness; weakened by heavy grazing; difficult to establish.

Orchardgrass. Cool season perennial; high-yielding with early spring and moderate midseason growth; excellent drought-tolerance and fair winter hardiness; coarse and unpalatable at maturity.

Reed canarygrass. Cool season perennial; high-yielding with early spring and good midseason growth; very good drought-tolerance and very good winter hardiness; poor palatability at maturity; difficult to establish.

Tall fescue. Perennial; moderate-yielding; good drought-tolerance and fair winter hardiness; poor palatability in warm summer months; weakened by heavy grazing.

Timothy. Cool season perennial; moderate-yielding with poor midseason growth; fair drought-tolerance and moderate winter hardiness; low palatability at maturity; weakened by heavy grazing and cutting.

Alfalfa, the queen. Long-lived perennial (except annual varieties); high-yielding with early spring and good midseason growth; good drought-tolerance, some varieties very winter hardy; cannot be grazed in seedling stage.

Alfalfa is the oldest crop grown solely for forage. It is native to the mountainous regions of southwestern Asia, in the vicinity of Iran and the Caucasus Mountains of southern Russia. It was grown in ancient times by the Arabians and Persians, and was then introduced into Europe and from there into Central and South America by the first Spanish explorers and settlers. Although grown to a small extent on the East Coast of the U.S. in the 1700s, alfalfa really succeeded in North America after seed was brought in the early 1840s by settlers sailing around Cape Horn to California. The name "alfalfa" comes from Arabic and means "best fodder." It is often called lucerne in other parts of the world.[3]

For anyone who feeds livestock, *the growing of high quality, healthful forages should be your number one concern.* **That is what will give the maximum production by your animals, as well as promoting their health and well-being. Also,** *high quality hay can be an excellent cash crop* **in many parts of the country.**

Let's see how *you* **can do it.**

_____

[3] D. Smith, *Forage Management in the North*, 1962, p. 74.

CHAPTER **2**

Care and Feeding

T he place to begin with growing really great alfalfa and other forages is at the beginning—with establishing the stand. If the plants do not get off to a good start, they will likely be sickly, have disease and pest problems, yield poorly, and the stand may die out quickly.

Soil. And the place to begin with establishing the stand is obviously the soil, because soil fertility and soil conditions play the major role in plant growth and crop yield—and most of all, *crop quality.* When you feed high quality forage to your livestock, they not only will produce more on the same quantity (or less) of feed, but they will also be healthier, and who can't do with lower vet bills and fewer dead animals?

What kind of soil, fertility, and soil conditions do alfalfa and other forage crops need to establish a good stand?

Needs. Alfalfa requires a well-drained soil for maximum production.[1] Soils two feet or more in depth are also necessary for best growth, since alfalfa is capable of developing a deep root system if root growth is unrestricted. Soils in which rooting depth is limited by either a shallow hardpan, a high water table (poor drainage), or bedrock are less suitable for alfalfa production.[2]

Hardpan. Have you ever dug up an alfalfa taproot and been surprised to find it bent at a right angle six or eight inches below the surface? This is dramatic evidence that a hardpan severely restricts root penetration, the use of deep nutrients, and therefore plant growth. The vast majority of farmland today is plagued by hardpans, as evidenced by water accumulating in low spots and even in high spots and on slopes after a rain. A hardpan not only restricts water penetration (and thereby increases water runoff, erosion, and flooding), but it also seals off the lower layer of soil from air. *Good soil aeration is vital for "healthy" soil,*

---

[1] M.B. Tesar & J.A. Jackobs, p. 418 in *Alfalfa Science and Technology*, 1972.

[2] *Alfalfa*, North Central Regional Extension Publication 113, p. 19.

because roots need oxygen and so do the beneficial soil microorganisms (bacteria, actinomycetes, and fungi), which are tremendously important for maintaining healthy soil and for growing healthy, high quality crops.

Aeration. In order for the soil to be well aerated (and to overcome a hardpan), the soil must be loose, spongy, and crumbly. In other words, it must have good structure or tilth. *Good soil structure can be obtained and maintained for the long run by having an adequate amount of humus* **and** *the beneficial soil organisms that produce it.* Humus is decomposed organic matter—the plant residues and manures which should be returned to the soil. When organic matter is worked into the upper layers of the soil, a "volunteer army" of bacteria, actinomycetes, fungi, and worms should be there waiting to attack it and convert it into dark, fine-textured, rich-smelling humus.

Humus. Abundant humus in soil provides many benefits:

1. It is a storehouse of essential plant nutrients (especially nitrogen, phosphorus, and sulfur) and growth-promoting substances (hormones and vitamins).

2. It helps make some nutrients more soluble and available to plants. Nutrients are released slowly throughout the growing season, as the plant needs them.

3. It contributes to good soil structure (tilth) by producing small crumbs (aggregates) of soil particles, allowing good air and water penetration. Water-holding capacity is also increased, and therefore drought resistance. Erosion from both water and wind is reduced. The soil is loose and easy to work.

4. It protects plants from diseases, pests, toxic chemicals, high salt levels, and drastic changes in pH (acidity/alkalinity).

What to do. "Soil organic matter is an important soil characteristic that improves tilth, water intake and water-holding capacity."[3] The usual measure of humus on laboratory soil tests is percent organic matter, although this does not distinguish between fresh, unrotted organic matter and true humus. By digging in your soil, you can see if last year's crop residues and manure are rotting quickly to form humus. If they are not, the problem may be due to "dead" soil without an adequate population of the humus-forming microorganisms (possibly because of toxic agricultural chemicals) or to tight, poorly aerated soil, causing anaerobic conditions (little or no oxygen). Compaction from use of heavy farm machinery is a contributing factor to anaerobic soil. Reducing or eliminating toxic

---

[3] D.A. Rohweder & J.M. Sund, *Establishing and Safeguarding Forage Seedings,* Univ. of Wisconsin-Extension Publication A1102,1974, p. 2.

chemicals and increasing humus content will alleviate these problems, but if your soil is so tight and "dead" that organic matter will not decompose quickly to form humus, then you can break out of this vicious circle by use of a soil conditioner to loosen soil and stimulate soil life. Depending on your soil's needs, some rock fertilizers can help condition soil (calcitic lime and soft rock or colloidal phosphate) or some commercial soil conditioners can be beneficial (although some kinds are not so helpful or can even do long-range harm). Inoculating the soil with beneficial bacteria and other organisms may help (if the soil conditions are already fairly good, and not toxic).

Desirable levels of "organic matter" on soil tests are from 2-5%, or even up to 10%, provided the soil is loose and "alive" with organisms.

A stubborn hardpan can be broken up by subsoiling or by plowing a little deeper each year, but a good earthworm population can do a better and quicker job of it.

Nutrients. The mineral elements that are most essential for good stand establishment are calcium (Ca), phosphorus (P), and potassium (K). Calcium is needed for cell division, cell wall formation, and root growth. Phosphorus is used for energy transfer and other metabolic functions in the plant, and also it increases root growth. Adequate phosphorus is especially critical for stand establishment.[4] Potassium is required to activate many cell enzymes and for food transport in the plant.

Nitrogen application in the nitrate form will help to establish alfalfa if your soil is low in nitrogen. The nitrogen-fixing bacteria which will later develop in legume root nodules require the trace elements molybdenum (Mo), cobalt (Co), iron (Fe), and copper (Cu). In properly fertilized soil with adequate humus and soil organisms, these trace elements should not be deficient, but the soils in some parts of the country *are* deficient in one or more trace elements, so some may have to be added. Be sure not to supply too much, because trace elements are only required in very small amounts, and some are toxic to plants or animals in too large amounts or in out-of-balance soil. Natural fertilizer sources such as manures and rock fertilizers can often supply trace element needs, and a good microorganism population will make them available to the plant.

Test first. But how can you know how much of what kinds of fertilizers to apply if you have no idea of what your soil needs? So before you do anything, you should have your soil tested by a reliable testing lab. Unfortunately, different soil testing labs differ in their testing methods and

---

[4] A.M. Decker, *et al*, in *Forages*, 3rd ed., 1973, p. 387.

interpretation of results, so you can send the same soil sample to two labs and get two different sets of numbers and fertilizer recommendations. Because of the prevailing beliefs about crop fertilization, most labs tend to recommend relatively too much potassium and too little calcium and phosphorus. The best soil testing methods for determining plant needs are those that test for readily available (soluble) nutrients (see Chapter 3).

Guidelines. It is impossible to give definite recommendations in this book without knowing what your soil needs, but the soil should have a high level of available calcium and phosphorus. If your soil needs these elements, good sources are calcite lime plus soft rock phosphate. These plus an application of organic matter (6 to 10 tons/acre of fresh cattle manure, or $1/2$ to $1/3$ that amount of poultry manure, or 1 to 3 tons/ acre of compost) will take care of most nutrient needs of alfalfa and other forages. The organic matter will provide enough potassium as long as calcium and phosphorus are high (see Chapter 3). Fresh organic matter should not be applied in excess nor be plowed in too deeply (below 5 to 8 inches) because it may not decompose properly, but may putrify and release toxins. It should be worked into the upper several inches (the aerobic zone).

The soft rock phosphate should be applied before or at the same time as the lime, since by itself, the lime tends to leach downward. They should not be plowed under deeply, and can be left on the surface.

Lime. If you live in a part of the country with low magnesium soils, dolomitic lime (calcium-magnesium carbonate) should still not be used; it has the disadvantage of being harder and slower to break down than calcitic lime, plus its high magnesium content can lead to tighter soil and nitrogen depletion if in excess. Calcitic limestone (calcite, calcium carbonate) has none of these disadvantages and should be used instead.

The more finely ground the lime is, the more rapidly it becomes available and the less that is needed. Mesh sizes of 90 - 99 or finer give almost "instant" availability, but they are hard to spread on windy days, and special spreaders may be needed. The old "E-Z Flow" and Gandy spreaders and the larger Stolzfus and Webster spreaders will handle fine lime.

pH. Standard recommendations state that alfalfa should have a soil pH of 6.5 to 7 or 7.5, which is above the average for most crops (6.2 - 6.8). Actually, not so much attention should be paid to the exact pH figure because (1) the pH of soil changes constantly, even from day to day, and (2) the pH readings produced by a soil testing lab depend on the methods used. For example, if the soil samples are finely ground before testing, the pH readings will be somewhat higher than under field conditions because small lumps of lime will be ground up and made more available.

Perhaps one reason a higher pH is recommended for alfalfa is that alfalfa requires high levels of calcium, and large amounts of lime are applied to raise pH, automatically supplying the crop's need for calcium.

Low pH (below 6.0) can have detrimental effects in reducing or eliminating growth of beneficial soil bacteria, including nitrogen-fixing bacteria, but high quality forage can be grown on acid soil, *provided* it has balanced and high fertility.

Seedbed. The best seedbed for forage establishment is firm and moist. Firmness will prevent loss of essential moisture; however, a crust is very detrimental to seedling emergence. Good tilth and humus content will prevent crusting. Fall plowing and spring disking and harrowing work well in most areas; however, fall plowing is not recommended in areas where erosion could be increased (steep slopes and high rainfall). Since shallow seed placement is necessary for good emergence, the use of a corrugated roller or packer will provide firmness.

Which seeding method? Whether you want to use broadcast, drill, or band seeding methods may depend mainly on your situation and available equipment. With good soil conditions, any seeding method can give good results. Under less than ideal conditions (low fertility or dry weather), band seeding (placing a band of seed directly over a band of fertilizer 1-2 inches deep) has been proven superior.[5]

Companion crops. In northern and eastern parts of the U.S., most alfalfa is sown with a companion crop (nurse crop) in spring seedings (not in summer or fall seedings). Besides providing an additional crop, companion crops protect the soil from erosion and keep out weeds before the alfalfa is established. However, companion crops can have disadvantages: they can compete with or inhibit the alfalfa seedlings by competing for light, moisture, and nutrients. Therefore, less leafy species or smaller seeding rates of companion crops should be used.

Commonly used companion crops are flax, peas, spring wheat, spring barley, and early maturing oats. Winter wheat, winter barley, winter rye, and late varieties of oats are poor companion crops for alfalfa.[6] Early mowing, grazing, or harvesting of small grain companion crops before the boot stage will help reduce competition with alfalfa.

The percentage of grass in legume-grass mixtures should generally be less than 25-40% (up to 50% in pastures) because too much grass will lower the protein content of the hay and may require more nitrogen than the legume can supply. Legume-grass mixtures that do

[5] M.B. Tesar & J.A. Jackobs, p. 424 in *Alfalfa Science and Technology*, 1972.

[6] *ibid.*, p. 425.

well together include (from Univ. of Wisconsin-Extension Publication A2906,1978, p. 4):

Mixture	Uses
alfalfa-bromegrass	hay, silage, green chop, pasture
alfalfa-orchardgrass	hay, silage, green chop, pasture
red clover-timothy	hay, silage, green chop, pasture
Ladino-orchardgrass	pasture, green chop, silage
white clover-bluegrass	pasture
trefoil-bluegrass or timothy	pasture

If no companion crop is used (direct seeding, clear seeding), weeds and erosion could be problems on poor soils. On steep slopes, a thin mulch of straw or manure will help reduce erosion. If the available phosphorus level of the soil is about twice as high as potassium, and if the soil is well aerated, weeds are not generally a problem. If you wish to use a herbicide for weed control, consider that most toxic chemicals tend to upset the soil's beneficial microorganism population, which can lead to humus depletion and lowered soil fertility. The use of a surfactant or wetting agent can allow you to greatly reduce the amounts of herbicides used.

Seeds. To get your forage crop off to the best start possible, use high quality (high test weight) seed and a suitable variety which is adapted to your climate. Yield, winter-hardiness, disease and pest resistance, and maturity time are factors to consider in choosing a variety. Details on varieties will be given in Chapter 6.

Inoculation. Legume seed should always be inoculated with the proper strain of nitrogen-fixing bacteria to insure development of root nodules. The extra cost is small, while the benefits are great. Pre-inoculated seed can be purchased or you can apply the inoculant at seeding time. Inoculant or inoculated seeds should be stored in cool temperatures (below 60°F in a refrigerator is fine) and used as soon as possible (not over six months after purchase).

Generally, seed treatment with fungicides is unnecessary for small-seeded legumes and grasses.

Planting depth. Optimal seeding depth for legumes and grasses is less than one inch. In fine-textured and moist soils, seeds should be planted closer to the surface, from $^1/_2$ to $^1/_4$ inch. In summer or drier periods or in sandy soils, deeper planting ($^3/_4$ to 1 inch) is recommended.

Seeding rate. There are several factors to consider regarding seeding rates:

1. *Moisture.* If the soil will not have much moisture later in the year (especially sandy soils), lower seeding rates will reduce competition for

moisture among the seedlings. Adequate humus will increase available soil moisture.

2. *Soil conditions.* Low soil fertility or acid soils will require higher seeding rates to insure that enough seedlings survive. Proper fertilization and adequate humus will overcome these problems.

3. *Species and variety.* Different grasses and legumes and their varieties differ in their germination rate, number of seeds per pound, and growth-form (some spread out in growth more than others). Some useful information is provided in the following table, from Iowa State University:

Forage seed characteristics

crop	weight per bushel (lbs.)	seeds per pound	seeds per sq. ft. (at 1 lb./acre)
alfalfa	60	200,000	5
alsike clover	60	700,000	16
birdsfoot trefoil	60	375,000	9
Ladino clover	60	800,000	18
red clover	60	275,000	6
sweetclover	60	260,000	6
bromegrass	14	136,000	3
Kentucky bluegrass	14	2,177,000	50
orchardgrass	14	654,000	15
reed canarygrass	30	533,000	12
tall fescue	24	227,000	5
timothy	45	1,230,000	28

University of Wisconsin recommendations for alfalfa seeding rates are 10-12 pounds of live, pure seed per acre for pure stands, 15 pounds per acre if quackgrass may be a problem, and 16-18 pounds per acre if you wish to harvest in the year of seeding.

Use the number of seeds per pound to figure seed mixtures. For example, it would take only about one-fifth the amount of orchardgrass seed to equal bromegrass.

Timing. The timing of stand establishment must be adjusted to your local climate and possible crop rotation schedule. In the North and Northeast, the best time is spring; otherwise dry summer weather may not allow enough growth to survive the winter (companion crops should not be used for late seedings because they compete with the legume and slow the establishment). In the South, late summer is the best time for seeding.

CHAPTER 3

Quality

Can you believe that you can take pretty much identical-looking hay from neighboring fields, feed 50 pounds a day from one field to a cow and have her drop in milk production and get sick, and feed half as much from the other field and have the cow rise in production and be healthy? What is the difference between the two samples of hay? *Quality!* And why the differences in quality? Because of the way the two farmers treated the two fields—the way they fertilized their soil, the way they managed their hay stands, timing of cutting, and perhaps other little tricks that can make a world of difference. The difference between only fair and excellent quality crops. And the difference between being a successful farmer or rancher and going bankrupt.

Until about 1960, alfalfa yields of 4 or 5 tons per acre were considered very good. In the 1970s, yields of 8 to 10 tons per acre were being obtained in the corn belt.[1] In the 1980s, a few farmers were raising over 12 tons per acre, and in southern California, with a long growing season, yields of over 30 tons per acre were reported. Increased use of fertilizer, herbicides, and insecticides, along with improved varieties, irrigation, and better cutting management were responsible.

Sounds great, doesn't it? Is that all it takes? But not every farmer can use irrigation or live where there is a long growing season. Not all farmers are using pesticides, because some farmers don't have forage disease, weed, and insect problems. What about the farmers that are growing the hay that cows only eat half as much of? What are they doing? How can *you* get excellent yields of high quality hay or silage?

Do you have to cut your alfalfa when it is only 8 inches or a foot high because it flowers too soon? Those farmers who practice the methods we are describing in this book are cutting alfalfa that grows over 40 or 50 inches tall before it blooms. And they are making hay

[1] C.L. Rhykerd & C.J. Overdahl, p. 437 in *Alfalfa Science and Technology*, 1972.

that's leafy and with succulent stems that are *solid in the center*, with no hollow air space. That's quality hay! The kind that cows love to eat, but don't eat 50 pounds per day of. Those farmers have so much hay left over that they can sell to their hard-up neighbors at premium prices. They often rent *less* land because of their bumper crops. If you would like to join them, it's not that mysterious or difficult. But it *does* require you to be willing to do things a little differently from your neighbors or from what the traditional advice-givers tell you to do. First a word about quality.

What is quality? Is a high quality hay crop one that produces the largest number of bales or tons per acre? Or one that produces the greatest amount of meat, milk, or wool per pound and that fosters good animal health and reproduction? I think you would agree that mere bulk does not insure high quality. After all, unless you are in the seed business, the reason you are growing a forage is to feed animals, and the results shown by the animals are what count.

But how is the quality of animal feed measured? There are many ways of measuring, or attempting to measure feed quality, and we do not need to go into detail about them here. Briefly, some methods attempt to measure the total energy value of feeds, based on actual feeding trials with animals or by measuring the number of calories contained in the food. Other methods measure only the protein content of the feed as an index of quality. In one such method, one of the most commonly used, only the nitrogen content of the feed is measured and the result is multiplied by 6.25 to give an *estimate* of protein content. This quantity, called crude protein, is a very inadequate and often misleading protein measure, for in poor quality sick plants, there is often an excess of nitrates and other non-protein forms of nitrogen. A better measure of protein quality would be an analysis of the amino acid content, since a certain balance of amino acids is needed for high quality protein, but such tests are very expensive.

Besides the protein and amino acid content of food, the mineral content is also vital in promoting good animal and human health, for many serious health problems and diseases can be cured with a diet containing sufficient minerals, in biologically useful forms. "Well-defined pathological symptoms appear in livestock deprived of certain needed elements, and thus there is abundant proof that the health and well-being of animals is directly dependent on the mineral content of the soil on which their food is grown."[2]

Other methods try to measure the digestibility of food, correlated with feeding trials and the percent of fiber in the plants, since in forages, as the amount of cellulose and fiber increases, the protein content decreases.

[2] F.A. Gilbert, *Mineral Nutrition and the Balance of Life*, 1957, p. 6.

Really tall alfalfa is not a myth. This field measured 44 inches and was just beginning to bloom.

Pioneers

Pioneering studies in the use of animals to test food quality were done by W.A. Albrecht and G.E. Smith in the early 1940s. It was found that forages grown on soils with unbalanced fertility (too much or too little of one or more elements) may have the same content of certain nutrients (phosphorus, calcium) as hay grown with balanced fertility, *as measured by standard laboratory procedures*. But when fed to animals (sheep, rabbits), the balanced-fertility feed produced about 50% more growth (weight gain) than unbalanced feed. Thus, there are important growth factors (vitamins, essential amino acids) that are not measured by ordinary lab tests, but which give the feed higher nutritional value to animals (*Soil Science Society of America Proceedings*, vol. 6, p. 252-258 (1941); vol. 7, p. 322-330 (1942)).

Of course, the most important measure of food quality is how *your* animals respond and what your expenses and income are—not only the animals' production of food or fiber, but also their health and reproductive success. "Thus, the total advantage of high-quality alfalfa goes beyond that indicated by digestible nutrient content and is compounded by a potential for being consumed at higher levels, a faster rate of digestibility, and perhaps a more efficient conversion of digested energy to produce energy."[3]

Refractometer. *But how can you measure quality while the crop is growing without sending in samples for an expensive and hard-to-interpret lab test?* One very simple and inexpensive method being used by a growing number of farmers is a refractometer, an optical instrument that measures percent sugars in the sap of a plant, which is correlated with the plant's food-producing efficiency (photosynthesis) and with ultimate food quality. Refractometers are routinely used in the food industry, by canneries, wineries, and breweries for example, to measure the quality of the fruits and vegetables they buy from farmers or of the foods and drinks they manufacture.

Using a refractometer to check the quality of your crops is easy. Simply squeeze a few drops of juice from the stems or leaves of the plants onto the glass prism of the refractometer, close the "lid," and look through the eyepiece. The sugar content is read on a numbered scale in

[3] R.F. Barnes & C.H. Gordon, p. 603 in *Alfalfa Science and Technology*, 1972.

Brix standards (% sucrose) for a number of crops

	poor	*average*	*good*	*excellent*
alfalfa	4	8	16	22
corn, stalk	4	8	14	20
corn, young	6	10	18	24
sweet corn	6	10	18	24
field peas	4	6	10	12
grains	6	10	14	18
sorghum	6	10	22	30

units called Brix (same as percent). By comparing with standard levels (see table) or past readings that you have made, you can see how your crops measure up that day.

Because of the difficulty in measuring feed usability and protein quality, there may not be a correlation between refractometer sugar readings and protein test figures *from a feed analysis lab*. Not all "high protein" feeds are truly nutritious for your animals. The best indicator to go by is the refractometer reading. A feed with a high protein figure but a low sugar reading will not be of high quality for your animals. It would be better to have a lower protein figure and high sugar.

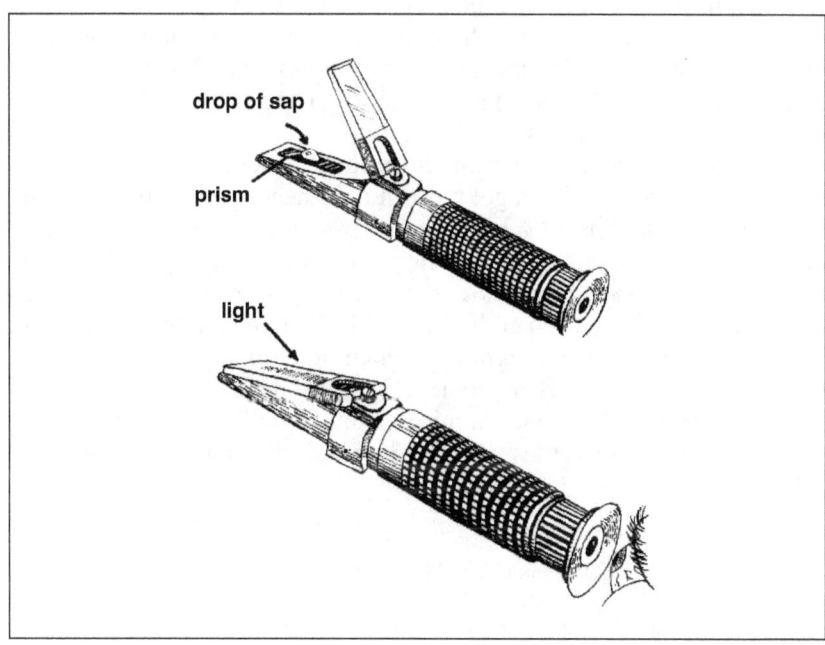

Growing high quality forages. Now that you have seen the importance of high quality feeds, how do you go about growing them yourself? If you have followed the suggestions in Chapter 2, you should have a vigorous stand of seedling plants off to a good start. What next?

Balanced fertility. The most important factor that determines the quality and yield of a crop is the soil. We have already covered the importance of loose, well-aerated soil with plenty of humus and beneficial microorganisms (Chapter 2). The other aspect of good soil is to have the right fertility—the proper amounts *and balance* of soil nutrients—what the plants need, when they need them. In Chapter 2, speaking of the elements needed to get a stand established, we mentioned several elements, especially calcium and phosphorus.

The actively growing forage plants need Ca and P plus the other elements needed by crop plants. According to a typical study (which may not have been made with balanced fertility), 10 tons of alfalfa removes from the soil 500 lb./acre of nitrogen, 50 pounds of phosphorus (120 pounds of P_2O_5), 500 pounds of potassium (600 pounds of K_2O), 350 pounds of calcium, 60 pounds of magnesium, and 50 pounds of sulfur.[4] Also needed are 11 oz./acre of boron, 10 oz. of zinc, 1.1 lb. of manganese, 1.5 lb. of iron, and 3 oz. of copper.[5] The iron and copper are also needed by nitrogen-fixing bacteria inside root nodules, plus they also need molybdenum and cobalt in very small amounts.

Major elements. As can be seen from the above paragraph, plants need more of some elements than others, so those are called the major elements. Nitrogen (N), phosphorus (P), potassium (K), and calcium (Ca) are the big four, with magnesium (Mg) and sulfur (S) sometimes called secondary elements.

Nodules. Nitrogen is definitely needed by legumes, but if things are working right, they can get most of their needs from nitrogen taken from the air (which is 78% nitrogen) by bacteria in root nodules. *However, if soil conditions do not allow the nodules to form or if they are not functioning well, legumes will take nitrogen from the soil,* just as any other plant that has no nodules. Healthy, active nodules will be pink or red inside due to a red pigment, leghemoglobin; a greenish color inside indicates a sick or dying nodule.[6] Nodule bacteria also require oxygen and freedom from toxic chemicals. Many alfalfa fields in this country have few or no nodules, mainly from poorly aerated soil and toxic substances.

---

[4] C.L. Rhykerd & C.J. Overdahl, p. 438 in *Alfalfa Science and Technology*, 1972.

[5] D. Ankerman & R. Large, *Soil and Plant Analysis*, undated, p. 62.

[6] J.C. Burton, p. 234 in *Alfalfa Science and Technology*, 1972.

Magnesium deficiency?

Some parts of the country that have magnesium-deficient soil include parts of western Oregon, Washington, and west-central California, the Atlantic coastal plain, and the Gulf Coast; also some irrigated sandy soils, and highly leached acid soils. Excessive use of high-potassium fertilizers, calcium sulfate, and some liming materials (oyster shells) can lead to magnesium deficiency (F.A. Gilbert, *Mineral Nutrition and the Balance of Life*, 1957, p. 83; R.L. Reid & G.A. Jung, p. 412 in *Forage Fertilization*, 1974).

Effective nodules can fix from 75 to 240 pounds of nitrogen per acre per year,[7] which isn't bad for free fertilizer! If your alfalfa doesn't have nodules, you had better find out what's wrong.

Calcium. Legumes are large consumers of calcium, and to obtain high-quality forage that promotes animal health and productivity, large amounts of calcium are essential.[8] *"Calcium serves so many important roles in the soil medium that it seems extremely doubtful that alfalfa culture can be very successful without it."*[9] Calcium is vital for cell division and healthy cell walls, for root growth and root hair formation, for enzyme activation, and protein production. A deficiency of calcium has been found to decrease resistance to insect pests.[10] Calcium also stimulates growth of beneficial soil microorganisms, including nitrogen-fixing bacteria, and helps counteract toxins in the soil and in the plant.[11]

In general, soils in the western U.S. are well supplied with calcium, but those in the East require additional calcium. High calcium lime (calcite) is preferred over dolomite lime, partly because of its more rapid availability and also because of its lower magnesium content, even in areas with magnesium-deficient soil. The fertilizer material sul-po-mag is a better source of magnesium, plus potassium and sulfur. You should have your soil tested to be sure, because there are localized areas in the East with excess calcium and calcium-deficient soils in the West.

If your soil needs it, calcium should be applied even though the pH

[7] *ibid.*, p. 242.

[8] R.F. Barnes & C.H. Gordon, p. 603 in *Alfalfa Science and Technology*, 1972.

[9] J.C. Burton, p. 240 in *Alfalfa Science and Technology*, 1972.

[10] E.L. Sorensen, M.C. Wilson, & G.R. Manglitz, p. 377 in *Alfalfa Science and Technology*, 1972.

[11] W.W. Woodhouse, Jr. & W.K. Griffith, p. 405 in *Forages*, 3rd ed., 1973; R.W. Pearson & C.S. Hoveland, p. 303-305 in *Forage Fertilization*, 1974.

The words of Dr. Frank Gilbert
are still up-to-date:

"Phosphorus, more than any other element, is the cause of low crop production, and the amount used on American farms is only a fraction of what is needed for the best results. The pastures and hayfields of the United States are especially in need of phosphorus, because most of them have never been fertilized, although cropped continuously for many years.

"It has been found that to obtain optimum yield of most crops, a much larger amount of phosphorus must be applied than is eventually removed in the first year's crop. This is because some of the phosphorus combines with the soil, that is, is 'fixed,' and is not immediately available.

"Of all mineral deficiencies in livestock, phosphorus deficiency is probably the most common and widespread.

"When there is even a slight lack of phosphorus in the diet, normal animal growth is prevented" (F.A. Gilbert, *Mineral Nutrition in Plants and Animals,* 1948, p. 14, 25, 29).

---

test results show no need for lime to "correct" soil acidity, because the plants need the calcium no matter what the pH. High pH can be caused by other elements than calcium.

"Calcium deficiency in the soil has come to be plant-nutrient problem number one in agricultural production . . . and calcium is now considered to be far more important in the production of food high in nutritive value than was formerly believed."[12]

Phosphorus. Phosphorus is the major element most often overlooked in crop fertilization programs. Even though the soil has large amounts of phosphorus in its minerals, it is nearly all chemically locked up and unavailable to the plant at any one time. If soil conditions are favorable, the beneficial soil microorganisms will slowly break down mineral phosphorus and make it available, but unfortunately, most of today's soils are in such poor shape that plants cannot obtain their phosphorous needs.

Phosphorus is essential for quality crops. "The lack of available phosphorus is usually reflected in low yields, poor quality, and delayed maturity. . . ."[13] Experiments with alfalfa have shown that adding

[12] F.A. Gilbert, *Mineral Nutrition and the Balance of Life,* 1957, p. 66.

[13] *ibid.,* p. 47.

[14] C.L. Rhykerd & C.J. Overdahl, p. 447 in *Alfalfa Science and Technology,* 1972.

What potassium can do

Experiments with different amounts of potassium fertilizer on alfalfa showed that feed mineral content (except for potassium and iron) decreased with increased potassium fertilization (D. Smith, *Agronomy Journal*, vol. 63, p. 497-500, 1971; R.L. Reid & G.A. Jung, p. 412-413, 415 in *Forage Fertilization*, 1974). One study showed that high potassium fertilization increased the nitrate level (potentially toxic to livestock) of pearl millet and Sudangrass, while adequate calcium and magnesium reduced nitrate (B.A. Schneider & N.A. Clark, *Agronomy Journal*, vol. 62, p. 474-477, 1970). Also, over-application of potassium can lead to "grass tetany" in animals because of reduced magnesium levels in the plant (R.L. Reid & G.A. Jung, p. 414-415 in *Forage Fertilization*, 1974).

phosphorus fertilizer can increase protein content.[14] Phosphorus (and calcium) deficiency in alfalfa decreases milk production.[15] Phosphorus is needed for all cell activities by transferring energy within the cell, it is part of the cell's genes and membranes, and is important in growing roots and stem tips. Phosphorus is especially needed by young plants; as much as 75% of a plant's supply of phosphorus may be absorbed by the time it has produced 25% of its dry weight.[16] Phosphorus also directly or indirectly increases drought and disease resistance and nitrogen fixation, and decreases maturity time, especially in cool temperatures.[17]

Although superphosphate fertilizers are sometimes recommended for forages, nearly all (80 - 90%) of the soluble phosphorus they contain "in the bag" reverts back in a matter of hours to insoluble forms that are unavailable to the plant.[18] Better forms of phosphorus are the rock phosphates, mainly soft rock phosphate, or colloidal phosphate, which is already in a form available to plants (hard rock phosphate is not available until soil life acts on it, a slow process). Basic slag can also be used as a less desirable phosphorus source.

Potassium. The main problem with growing alfalfa today is that traditional recommendations are for too much potassium in proportion to the other elements, and you cannot grow really top quality forage when the soil's available potassium exceeds available phosphorus or

[15] R.F. Barnes & C.H. Gordon, p. 603 in *Alfalfa Science and Technology*, 1972.

[16] L.F. Seatz & C.O. Stanberry, p. 155-187 in *Fertilizer Technology and Use*, 1963.

[17] *Phosphorus for Agriculture: A Situation Analysis*, 1978, p. 42-111.

[18] *Profitable Management of Wisconsin Soils*, 1972, p. 35.

Calcium–Potassium

● From W.A. Albrecht,"Soil Fertility as a Pattern of Possible Deficiencies," *Journal of the American Academy of Applied Nutrition*, vol. 1, 1947, p. 17-28:

"The calcium-potassium ratio . . . has given us a pattern of the protein possibilities in the crop. If nature, under less rainfall, has left much calcium in the soil, we have a proteinaceous crop. If the soil is under higher rainfall to give a small amount of calcium in relation to potash, then we have a carbonaceous crop. The validity of this belief, namely that a liberal supply of calcium in the soil in relation to potassium represents production of crops rich in proteins and minerals, while the reverse relation give crops high in carbohydrates—thereby low in proteins and minerals—was tested. Soybeans were grown with increasing amounts of potassium available in the soil and associated with constant amounts of calcium. Three ratios of calcium in potassium were used while all other nutrients were liberally supplied. Increasing the potassium increased the forage yield to a maximum of 25%. This fact would draw ready applause for an experimenting agronomist. Such work can win funds in support of it as research. But the buffalo of the western plains didn't evaluate herbage in terms of bulk. Our livestock does not use that criterion either. Hence, while increase of bulk may appear laudable, fixing our attention on bulk in relation to soil fertility has been leading us to grow more crops with serious deficiencies as feeds.

"Chemical analyses were made of the forage. The nitrogen content of the smallest of the three crops was 2.8%; of the intermediate crop 2.5%; and of the largest crop 2.19%. While we increased the bulk 25%, we reduced the concentration of nitrogen, and therefore the protein, by more

continued . . .

Calcium–Potassium, continued

calcium. Now don't get me wrong, you can grow lots of alfalfa with high potassium—and pretty good looking alfalfa, too. Scores of research experiments prove that.[19] But it is not the highest quality animal food. It is what is called *carbonaceous*—too much cellulose. Plenty of stiff stems. What animals need is *proteinaceous* feed, with an increased proportion of high quality protein, and that kind of forage can only be grown when the soil has plenty of calcium and phosphorus and not too much potassium (see box). "The luxuriance of a plant is not necessar-

[19] C.L. Rhykerd & C.J. Overdahl, p. 448, 453 in *Alfalfa Science and Technology*, 1972.

[20] F.A. Gilbert, *Mineral Nutrition and the Balance of Life*, 1957, p. 6.

than that figure. So that the greater amount of total protein was not in the largest but in the smallest crop.

"The phosphorus concentration, by analysis, was .25% in the small crop; .18% in the intermediate; and .14% in the largest crop. Assuming that the cow could digest it completely, she would be compelled to eat approximately twice as much of the larger crop to get the same amount of phosphorus. In the case of the calcium, this was approximately .75% of the dry weight in the smallest crop and only .27% or about one-third as much in the largest crop. Can any cow increase her consuming capacity by three times? We can't expect her to become a hay baler.

"We need to be concerned not only with the bulk of the crop, but also with the synthetic operations of the plant in using the fertility elements from the soil to convert the carbonaceousness over into proteinaceousness. Those processes make "grow" foods instead of merely "go" foods. They must be more generally appreciated if we are not to invite nutritional deficiencies more commonly."

• From Dr. Frank A. Gilbert, in his book, *Mineral Nutrition of Plants and Animals* (1948), p. 43, 45:

"The present opinion seems to be that a high potassium content of the soil when not balanced by other essential elements, especially calcium, tends to produce a plant especially high in carbohydrates. A normal amount when calcium and phosphorus are deficient will have the same effect and will produce plants of a very low biological value with a lesser change in caloric value. On the other hand, a more nutritious proteinaceous and less carbonaceous plant grows when the potassium-calcium ratio is weighted in favor of the latter element."

ily a criterion of its mineral content, and the usual commercial fertilizer may or may not add to a deficient soil all that is needed."[20]

For growing top quality forages, the soil's available calcium should be higher than available phosphorus, and phosphorus should be higher than available potassium. Another advantage of maintaining a high phosphorus - low potassium ratio is that weed problems will be greatly reduced or eliminated. Also, because the plants will be healthier, disease and pest problems will be minimal (see Chapter 5).

By "available," we mean readily available to the plant, or water soluble. Unfortunately, most soil testing labs use methods that give too high readings for most elements and thus not enough fertilizers may be recommended (on most soil tests, however, the P_1 test does approximately equal the amount of phosphorus readily available to the plant).

Because of over-application of potassium fertilizers (potash) in the past and the large supply already present in most soils (except sand), most fields need little if any additional potassium to grow high quality forages. If after several years some is needed, manures or crop residues should supply plenty. If even then some is needed, be sure to use potassium sulfate rather than the commonly sold potassium chloride (muriate of potash, kalium potash), since chloride is detrimental to crop quality and soil, especially at high rates. It has been shown to "burn" new seedings of alfalfa.[21]

Secondary elements. Smaller amounts of magnesium and sulfur (in the sulfate form) are needed than of the previous four elements. Magnesium is needed as a part of the plant's chlorophyll and is required for many enzyme and metabolic activities within the cell. Sulfur is a necessary part of some of the essential amino acids (cysteine and methionine) as well as having several other metabolic functions in the plant.

In most soils, magnesium and sulfur are not in short supply; in fact, "acid rain" delivers a more than adequate supply of sulfur (50 pounds/acre or more) in many areas. Nevertheless, some regions may have soil deficiencies in these elements, so a soil test is important to be sure. If magnesium is needed, dolomitic lime is a cheap source, but not a good source, since it is very slow to break down. If sulfur is needed, be sure to use a sulfate form, such as calcium sulfate (gypsum), rather than elemental sulfur (flowers of sulfur), which has harmful effects on the soil and is slow to become available to the plant. Sul-po-mag supplies both sulfur and magnesium, as well as potassium.

Trace elements. The remaining essential nutrients are only needed in tiny or trace amounts, but they are just as important as the major and secondary elements. The ones most needed by alfalfa are boron, zinc, manganese, iron, and copper, plus molybdenum and cobalt for nitrogen-fixing bacteria. Most soils have adequate supplies of all of them with the possible exception of boron, which is often deficient in U.S. soils in the entire eastern one-half and in the Northwest and California. A boron shortage can seriously reduce alfalfa yield. Boron is needed for cell division, normal growth and maturity, for protein and seed production, and other metabolic cell activities. Boron deficiency problems in alfalfa can be worsened by high potassium and high pH soils (neutral or above).[22]

[21] C.L. Rhykerd & C.J. Overdahl, p. 462 in *Alfalfa Science and Technology*, 1972; D. Smith, *Agronomy Journal*, vol. 63, p. 497-500, 1971; S.A. Barber, R.D. Munson, & W.B. Dancy, p. 331 in *Fertilizer Technology & Use*, 2nd ed., 1971.

[22] F.A. Gilbert, *Mineral Nutrition and the Balance of Life*, 1957, p. 170, 173.

Boron can most cheaply be supplied by applying borax, at whatever rate your soil requires (have a soil test run).

If a soil test shows any of the other trace elements to be deficient, apply recommended rates, but first be sure that the major elements are at adequate levels, especially calcium and phosphorus. In fact, once those elements are adequate, trace element "shortages" may disappear, because calcium stimulates soil microorganisms to release trace elements, and phosphorus is needed by the plant to transport them.

Balance. No single element is more important than another for optimal plant growth; all are necessary, but in different amounts or proportions. Therefore, the proper *balance of nutrients* is the important thing. And that means the nutrients *available to the plant*, not the total present (but tied up) in the soil. That's the only way to grow top quality crops.

One of the keys to keeping the right balance of nutrients is to have well-aerated soil with plenty of humus, for humus and its associated microorganisms tend to provide balanced nutrition to plants. But that's not all. They also provide many other valuable pluses. They produce growth-stimulating substances, improve the general vigor and health of the plant, and combat plant diseases and pests. Properly fertilized alfalfa stands can live 30 or 40 years.

Experiments have shown that unbalanced fertility may result in increased yields (bulk), but also give crops of lower nutritional value to animals. For example, alfalfa was fed to Guinea pigs. Moderately fertilized alfalfa increased the animals' weight gain, while alfalfa fertilized with too much phosphorus and potassium decreased weight gains.[23]

[23] W.F. Wedin, A.W. Burger, & H.L. Ahlgren, *Agronomy Journal*, vol. 48, p. 147-152, 1956.

CHAPTER 4

Harvesting

Now that you have a good stand of forage established and understand some principles of fertilization to obtain high quality, how can the stand be best managed to give optimal yields of the highest quality feed?

When to cut. There have been various ways used over the years to tell when to cut alfalfa, such as cutting at the bud stage, or at 10% bloom or half bloom, or full bloom, or when new shoots appear from the crown, or even by the calendar. The latter is obviously unreliable because of variable weather conditions, but when *should* alfalfa be cut? There are scientific studies which show that the plant's content of minerals and digestible nutrients is highest during the succulent growth stage and declines during flowering and maturity, while fiber content increases. Feed value is said to decrease.[1]

These studies were probably done with forages that were fertilized with too much potassium and perhaps other imbalances, because on-the-farm experience and certified production records show that top quality feed can be produced when alfalfa is not cut until early bloom (25 - 50% bloom). Cows can eat two-thirds to one-half as much of this kind of hay and still increase in production. Early-cut hay is often too high in nitrates and low in high quality protein.

Traditional studies have shown that the long-range greatest yields and best recovery growth are produced when alfalfa is cut at full bloom, although in the Southwest, greater yields may be obtained if cutting is done at 10% bloom (first bloom), but in the hottest weather stand density and yield will be reduced.[2]

The best method to judge when to cut is to use a refractometer (see Chapter 3) to measure the sugar content of plant juice every day

---

[1] R.F. Barnes & C.H. Gordon, p. 603-607 in *Alfalfa Science and Technology*, 1972.

[2] D. Smith, p. 486, 492 in *Alfalfa Science and Technology*, 1972, and p. 93 in *Forage Management in the North*, 1962.

or two as blooms begin to appear. When the sugar readings reach a peak or begin to level off, that is the time to cut. That is generally at 25 - 50% bloom. The appearance of new shoots from the crown is another sign to watch for, but is not as reliable as bloom stage or the use of a refractometer.

Signs of quality. *When you begin to get your soil fertility in the proper balance as described in Chapter 3 (high calcium and phosphorus, low potassium), you will be amazed that your alfalfa doesn't bloom at a height of 8 inches or a foot like it used to, but it will keep on growing—and growing—and growing. It may reach heights of over 40 or 50 inches before it is ready to cut. And it will be so thick that you may have to get a new heavy-duty mower! You won't get as many cuttings as your neighbor does, but you will get far superior quality feed and probably as good or better total yields.*

If you cut open several stems with a knife, you may begin to see the growth of "solid stem alfalfa," in which the entire stem is filled with succulent cells, not air.

Good quality hay will dry rapidly after cutting (because its cells contain more minerals and nutrients and less water) and can even be baled so wet that ordinary hay would heat up and burn the barn down.

Cutting. Cutting should be done with a sickle-bar or cutter-bar mower, which gives a good, clean cut to the stems. High quality hay will dry quickly and does not need to be crushed or crimped; in fact, this torture treatment can cause loss of nutrients from the crushed tissues. Cutting should always be made above any new shoots that are sprouting from the crown, since the growing point of legumes is at the stem tips, and cutting them will severely retard the next growth. This is not true of grasses, whose growing point is at or below ground level.

Drying aids. If you want to speed drying, there are sprays that can be used. Some are designed for poor quality forage and are basically a salt solution, but there are also ways of speeding drying and increasing the feed value of hay by spraying a carbohydrate solution before cutting.

Stimulate regrowth. If the soil needs any nitrogen, a nitrate-containing fertilizer can be topdressed at a low rate after a cutting to stimulate regrowth. Fertilizers containing ammonium nitrogen should not be used because ammonium will stimulate early flowering rather than vegetative growth (leaves and stems).

Also, if you did not apply the recommended lime and soft rock phosphate before seeding or in the fall or spring, they can be topdressed after a cutting.

Early cuttings. The first cutting of a new stand should be delayed

Alfalfa flowers

until the plants are strong and vigorous and have a good root system, generally 70 to 90 days after germination for a spring seeding.[3] Also, in the North the first cutting should not be made at an early stage of growth or the plants will be injured from low food reserves.[4]

Late cuttings. In more northern regions especially, alfalfa should not be cut or grazed during the period of 4 to 6 weeks before the first killing frost, or approximately between the first week of September and mid-October.[5] This is so the plants can store up enough food reserves to survive the winter. A cutting can be made after a killing frost (not a light frost).

Grow your own seed. Once you have good soil and a good, healthy stand, there is no reason for not setting a small plot aside for growing your own seed, adapted to your own soil and climate. To produce high quality seed, the soil needs a higher level of ammonium nitrogen (after the plants have attained some growth) and an adequate level of manganese. Bees are also necessary for legume seed production (not grasses). Either honeybees or wild bees (bumblebees, leaf-cutter bees, and alkali bees) can do the job of pollinating the flowers. Hives of honeybees can be rented from honey producers if there are not enough in your area.

[3] *Alfalfa*, North Central Regional Extension Publication 113, p. 21.

[4] D. Smith, p. 488 in *Alfalfa Science and Technology*, 1972.

[5] D. Smith, *Forage Management in the North*, 1962, p. 95.

CHAPTER 5

Problems

As you well know, not everything always goes according to plan. Problems sometimes develop. It is hoped that we can overcome them and use them as a learning experience, and not be overcome by them.

What should happen. Here's the way it *should* work. If we have the proper balance of soil fertility, well-aerated soil with plenty of humus and beneficial microorganisms, and if the weather is ideal, then we will have excellent growth of vigorous, healthy plants which will produce high yields of nutritious food and *which will resist diseases and pests.* Sounds like an unrealistic dream, doesn't it? But the only thing you have no control over is the weather, and if the soil is in good shape, even halfway poor weather will not stop you from getting good crops (a hurricane or a week of rain may set you back a little!).

Bugs. Believe it or not, a healthy, vigorous plant can ward off or resist the attack of infectious diseases and pests. This is a known fact that is being supported by more and more research, although the chemical companies may not want you to know about it.

Of course, it makes good sense to plant varieties of crops that have genetic resistance to certain pests and diseases, at least until you do get your soil in good condition (varieties will be covered in Chapter 6).

A large book could be written about all of the various diseases and pests that afflict alfalfa and other forages. They range from verticillium wilt to Texas root rot to alfalfa weevils to spotted alfalfa aphids. The book *Alfalfa Science and Technology* (1972) lists 3 bacterial, 24 fungal, and 3 viral and mycoplasmic diseases; 66 kinds of insects and mites; and 9 nematodes that attack alfalfa.

Pity the poor plant pathologists and entomologists who spend all their time looking at sick plants, identifying the critter that's causing the damage, and then using toxic rescue chemistry to try to save the crop.

The diseases and bugs are there for a reason—to clean up the unfit and unhealthy, and to tell us something's wrong. The best approach to diseases and pests is to *prevent* them, to build up your soil fertility so that crops will be vigorous and healthy, able to resist attack. It's just like human or animal health. Even though we are exposed to germs every day, we don't get sick unless we let our natural defenses down and don't get the right food or enough sleep or exercise. Also, *stress* can make you sick. The same is true of plants. They can be weakened by stresses—whether from unbalanced soil fertility, toxic substances, waterlogged soil, or adverse weather—and soon fall prey to diseases and/or pests.

Many studies prove this. For example, an excellent summary article by J.L. Dodd in the June 1980 issue of *Plant Disease*, is entitled "The Role of Plant Stresses in Development of Corn Stalk Rots." He points out that stalk rot fungi do not attack corn plants until they are under some kind of environmental stress or are of a susceptible variety. Corn has been found to produce chemical substances that inhibit corn borer larvae,[1] and an anti-fungus chemical has been found in wheat and corn.[2] A study by S.D. Kindler and R. Staples found that two alfalfa varieties were more susceptible to the spotted alfalfa aphid when the soil fertility was out of balance, either too little calcium or potassium, or too much magnesium or nitrogen.[3]

Crisis management. However, if your soil is not yet in good condition, or if the weather isn't cooperating, and your plants are under stress and are attacked by a disease or pest, all the Utopian theories in the world will not help. You have to save that crop! So go ahead and follow the advice of your local extension agent or whoever, but if you have the guts, don't apply the recommended insecticide or fungicide at the recommended rate. Try reducing the rate by one-third or even one-half. Usually the recommended rates are plenty high. Also, since only a small fraction of sprays actually reach their intended target, the use of a surfactant or wetting agent will help the pesticide penetrate more readily and allow you to use less poison. If you don't want to gamble on a whole field, at least try a test strip at reduced rates.

Remember that most pesticides upset something—either the plant's functions or the soil's balance of microorganisms, or something. There are usually kickbacks which you don't need. They only make it harder to get your soil in healthy condition later. So use as little toxic chemicals as possible.

[1] S. Beck, *Journal of Insect Physiology*, vol. 1, p. 158-177, 1957.

[2] A.I. Virtanen, P.K. Hietala, & Ö.U. Wahlroos, *Suomen Kemistilehti*, vol. 29, no. 1B, p. 143, 1956.

[3] *Journal of Economic Entomology*, vol. 63, p. 938-940, 1970.

Weeds. The same thing goes for weeds. Nearly all weeds grow best in unbalanced soils, especially those that are poorly aerated (anaerobic) or that have too much potassium. Velvetleaf (buttonweed), quackgrass, and bindweed prefer anaerobic soil.[4] Some herbicides (and other toxic chemicals) even make weed problems worse. For example, herbicidal control of foxtail encourages fall panicum.[5]

Poor quality forage species (poverty grass, beard grass) tend to take over land depleted in phosphorus and other minerals. When phosphate fertilizers are applied, the better quality plants (clover, bluegrass) tend to crowd out the poorer ones.[6]

Most weeds actually do not grow well in good, healthy soil and in competition with vigorous, healthy crop plants.[7]

So obviously, as with diseases and pests, unless toxic chemicals are necessary, the best way to attack the weed problem is to get your soil into good condition and grow vigorous, healthy crop plants.

Non-infectious diseases. There are a number of problems which are not caused by disease germs or pests. They are sometimes called abiotic, nonparasitic, or physiogenic diseases. Some are beyond the farmer's control, but others can easily be prevented by proper soil conditions and fertility. They include weather factors, nutrient deficiencies and toxicities, pollution, and anaerobic soil conditions.

Weather. Frost injury can occur to tops in spring or fall. High quality, healthy plants with a high concentration of sugar in their cells are more frost resistant than weak plants.

Frost heaving from alternate freezing and thawing can lift the taproots and break side roots. These injuries allow diseases to get a start. Damage most often occurs in waterlogged, hardpan soils, so having good tilth will go a long ways toward preventing this problem (see Chapter 2). The shallow roots of a companion crop also help.

Winter kill or winter injury can occur from the stand being covered by an ice sheet which suffocates it or from exposure to cold temperatures and dry air, especially in non-winter hardy varieties. Dark internal discoloration of the taproots and crown (root rot, crown rot) are the symptoms. Sudden temperature changes, such as an untimely freeze or a warm period during the winter can also kill or injure plants because they have lost their cold resistance, or hardening. The ability of plants to survive the winter (other than the genetically controlled factors found in the different varieties) is largely determined by their state of health and amount of food

[4] C. Walters, Jr. & C.J. Fenzau, *Eco-Farm*, 1979, p. 300, 345.

[5] *ibid.*, p. 308.

[6] F.A. Gilbert, *Mineral Nutrition and the Balance of Life*, 1957, p. 39.

[7] A.S. Crafts & W.W. Robbins, *Weed Control*, 3rd ed., 1962, p. 93.

stored in the roots during the fall. Obviously, good fertile, healthy soil is an essential factor in preventing winter kill.

High temperatures in the Southwest (and excessive irrigation) can also lead to root rot symptoms.

Nutrient deficiencies and toxicities. As we have said, a proper balance of soil nutrients is necessary for good health. Too little or too much of any nutrient or fertilizer material can cause problems.

Nutrient deficiencies can be produced in a number of ways. The soil may simply be deficient in one or more nutrients. Heavy rains can cause leaching of certain nutrients, such as nitrate or sulfate. Crop removal without organic matter recycling can deplete certain soil nutrients. Or there may be an imbalance; too much of one element can cause a deficiency of another; for example, high potassium can cause a boron deficiency,[8] and high magnesium can cause deficiencies in nitrogen, calcium, phosphorus, and potassium.[9] Too low (acid) or too high (alkaline) soil pH can change the availability of some elements. Low pH can decrease availability of phosphorus, nitrogen, potassium, sulfur, calcium, and magnesium, while high pH can decrease availability of phosphorus, nitrogen, magnesium, iron, manganese, boron, copper, and zinc.[10] Too little soil moisture and too much (waterlogged soil) interfere with the root's ability to absorb nutrients. Too low or too high temperatures can also do the same.

Toxicities of magnesium, manganese, aluminum, and ammonia can occur if they are present in too high amounts.

Others. Crop health and growth can also be damaged by air pollution ("smog," sulfur dioxide, ozone), acid rain, and waterlogged or tight or crusted soil, which leads to anaerobic soil conditions (no oxygen). Anaerobic soil damages roots from lack of oxygen and also causes anaerobic bacteria to release toxins. Damaged roots often react by developing gummosis, a sort of "hardening of the arteries" in which vital water and food-conducting vessels become plugged by gums and other secretions.[11]

How to tell. It is sometimes difficult to tell whether your crops are at peak health or are "slightly ill." Severe sickness is usually easy to spot because of visible discoloration or dying of above-ground parts. The following identification chart may be of value in determining severe deficiencies (however, some leaf symptoms are general and can have other causes, such as infectious diseases, air pollution, or anaerobic soil.

[8] *Micronutrients in Agriculture*, 1972, p. 256.

[9] C. Walters, Jr. & C.J. Fenzau, *Eco-Farm*, 1979, p. 181, 184, 192.

[10] H.D. Foth & L.M. Turk, *Fundamentals of Soil Science*, 5th ed., 1972, p. 186.

[11] R.D. Durbin, p. 102-110 in *Water Deficits and Plant Growth*, vol. V, 1978.

Also, you may have more than one element being deficient at a time). To use this chart, start with the Roman numerals (I and II), decide which of the two statements your plants fit, then go on to the capital letters (A and B) *under that Roman numeral.* Decide which letter your plants fit, then go on to the Arabic numerals (1, 2, etc.) *under that capital letter.* (Adapted from *Alfalfa*, North Central Regional Extension Publication 113, p. 26-27).

Identification of Legume Nutrient Deficiencies

I. Symptoms visible throughout entire plant, or on older (lower) leaves.
   A. Symptoms general throughout entire plant, with yellowing or drying up of lower leaves.
      1. Plants light green. Lower leaves affected first, then others; leaves turn pale yellow, then brown. Growth stunted .... **nitrogen**
      2. Plants dark green, but spindly and stunted. Leaflets and petioles ("leaf-stems") are tilted upward. Stems may turn red .... **phosphorus**
      3. Plants light green. Lower leaves die and drop off prematurely .... **molybdenum**
   B. Symptoms localized, with mottling (irregular spotting) or chlorosis (yellow to white to gray color) on lower leaves, but little drying up of lower leaves. Possible dead spots on lower leaves.
      1. Leaves turn pale green, then yellow between veins (base of leaf not affected). In later stages of growth, leaf edges curl downward and turn yellow, then bronze .... **magnesium**
      2. Yellow mottling (irregular spotting) or small white spots appearing around edges of leaves, then yellow border forming which later dies and falls off. Growth stunted .... **potassium**
      3. Brown spots and yellowing of tissue between leaf veins, later dying and falling out. Growth stunted .... **zinc**
II. Symptoms confined to younger (upper) leaves; growth stunted.
   A. Tip bud dies; tips or bases of young leaves distorted.
      1. Leaves near tip yellowed or sometimes reddish. Lower leaves a healthy green. Internodes (stem between leaf-stem attachments) shortened, giving a wreath-like appearance of leaves. Buds white or light brown. Little flowering .... **boron**
      2. Delayed emergence of main leaves. Leaves cup-shaped. Main leaves with dying areas, and yellow or whitish bands around other leaves. Tip buds deteriorate and petioles ("leaf-stems") break down ....
      calcium
   B. Tip buds remain alive.
      1. Leaves light green to yellow but veins darker green. Spots of dead tissue appear on leaves .... **man-ganese**
      2. Leaves yellow to whitish with veins green. Spots of dead tissue appear, especially at leaf edges, later falling away .... **iron**
      3. Leaves, including veins turning light green or yellow. Young

leaves affected first .... **sulfur**
4. Young leaves may wilt and wither without first turning yellow.
Excessive leaf shedding possible .... **copper**

Less severe deficiencies and toxicities are harder to diagnose. Often the plants look healthy, but growth may be slowed, yield may be reduced, and feed quality will probably be decreased.

Tissue analysis. Tissue analysis may be useful for spotting slight nutrient deficiencies, although it only tells you what nutrients are present in the plant tops, not the roots, and it doesn't tell you what might be the actual cause of the deficiency, nor whether the elements are in the biologically useful forms that produce healthy plants. If you want to send a sample to a testing lab, collect the top six inches from several stems at several different locations in the field when the plants are at the pre-bloom stage, dry the sample, and take or send it to the lab in a clean paper sack or other container (not an air-tight plastic sack; plants may mold). Tissue analysis results can be misleading for some elements as far as trying to diagnose soil deficiencies, because sometimes an element will be concentrated in a plant's tissues even when the soil is deficient. For example, special root fungi called mycorrhizae can supply adequate phosphorus to a plant growing in low-phosphorus soil.

Refractometer. Another tool that is easy and quick to use to check plant health is the refractometer. Its use for testing quality was covered in Chapter 3. By measuring the sugar content of your plants at frequent intervals or in different fields, and keeping accurate records, you may be able to spot problems. Remember that the sugar content reflects the plant's food-making activities of photosynthesis, and that photosynthesis is slower in cool or cloudy weather, and at morning and evening hours, so take your readings at approximately the same conditions and time every day. Also, the sugar content of the above-ground parts of forages fluctuates throughout the year, being high at maximum growth stage and low after cutting and in early spring.

Soil testing. Still another tool to spot problems is frequent soil testing, provided the testing methods measure readily available nutrients, not just the totals locked up in unavailable forms. Soil nutrient levels and pH can change considerably throughout the growing season, and frequent tests (two or three per year) can give you a better idea of what is going on in your soil, as well as help you spot problems.

What to do. Just about all the problems we have covered in this chapter can be prevented or alleviated by having good, healthy, well aerated soil. If the major nutrients (calcium, phosphorus, potassium, nitrogen) are at proper levels, the nitrogen-fixing bacteria and all the other beneficial soil microorganisms are "doing their thing," and there are no

Can you spot the alfalfa among the weeds?

droughts, floods, or summer frosts, your plants should be vigorous and healthy and be able to ward off diseases and pests, and your soil should not be plagued by weeds and nutrient imbalances or toxicities.

In an emergency, trace elements and some major elements can be supplied to plants through the leaves by foliar feeding, spraying a liquid solution of nutrients.

Do You Need GM Alfalfa?

After developing genetically modified (GM) varieties of corn, soybeans and cotton, which have been adopted by a majority of U.S. growers,* biotech companies have turned their attention to other crops, recently including Roundup Ready Alfalfa, developed by Monsanto and Forage Genetics.

"Roundup Ready," as with similar traits in other GM crops, means that the crop plants are resistant to the general-purpose herbicide glyphosate (brand name Roundup). This means the farmer can spray the field with glyphosate to kill most weeds (except those that have developed their own glyphosate resistance!) without killing the crop.

This feature may sound like a great advantage, especially in corn, soybean and cotton fields, but well-managed alfalfa fields rarely have serious weed problems. The dense alfalfa foliage shades out weeds, frequent cutting of alfalfa tends to prevent weeds from going to seed, and if the soil nutrient balance and organic matter are kept optimal, GM alfalfa just isn't necessary.

Another major problem with GM crops is the great possibility that the genetic traits inserted by the company can transfer (by pollen) to non-GM crops (or even to weed species), contaminating those fields. Organic and many sustainable farmers DO NOT want their crops contaminated by GM seed, and must pay for expensive crop quality tests if that is a possibility. Many consumers do not want to eat GM food either. GM crops are supposed to be surrounded by a buffer zone to prevent pollen transfer, but contamination incidents have already occurred.

Then there are possible intimidating legal entanglements that can occur when GM crops are grown. The company has patented the GM crop variety, and requires customers to sign a contract or agreement which brings very stiff penalties if the farmer saves GM seed for replanting. Also, the farmer bears legal responsibilities if his GM crop contaminates neighboring fields.

After Monsanto began to introduce Roundup Ready Alfalfa, with the USDA's approval in 2004, an environmentally conscious organization brought a lawsuit and in 2007 a California federal judge declared the USDA's approval null and void because proper environmental impact procedures were not followed. So for now, Monsanto cannot market this unnecessary product.

---

* This ready adoption of GM crops by American farmers is in spite of the fact that many foreign countries do not want them and research indicating that animals fed GM food develop serious health problems, including poor growth, and reproductive abnormalities.

CHAPTER 6

Varieties

There are hundreds of varieties of alfalfa and other forage species that have been developed by plant breeders. Varieties are developed to have various desirable characteristics: winter hardiness, disease and pest resistance, high yield, and early maturity time. As we have seen in previous chapters, good soil and healthy plants can overcome many of the problems that certain varieties are developed for. Nevertheless, it makes good sense to use good quality seed and to choose varieties that are suited to your climate and that have pest and disease resistance, since your soil conditions may not be ideal, or adverse weather may put plants under stress, making them more susceptible to problems.

Many new varieties are constantly being developed and tested. If they prove successful they eventually become "standard" varieties. Some of the recently developed varieties include spreading or creeping alfalfas that develop new crowns when the old ones are damaged or diseased (they do well on steep hillsides and poor soil, and have good winter hardiness), and varieties that fix more nitrogen than previous varieties.[1]

A few of the best and most popular varieties are listed below (information from *Alfalfa Science and Technology, 1972; Forage Management in the North*, 1962; Univ. of Wisconsin-Extension Publ. A1525 and A1525-1; and *Forages*, 3rd ed., 1973):

Alfalfa
- suitable for northern U.S.
  Vernal: winter hardy, wilt resistant, high yielding
  Iroquois: winter hardy, wilt resistant, high yielding
  Agate: winter hardy, resistant to wilt and phytophthora root rot

---

[1] F. Zahradnik, *The New Farm*, January 1983, p. 31, 33.

- suitable for South and Central U.S.
  Kanza and Cody: wilt and aphid resistant, moderately winter hardy
  Buffalo: wilt resistant, moderately winter hardy
- suitable for the West and dry climates
  Vernal: winter hardy, wilt resistant, high yielding
  Ranger: winter hardy, wilt resistant, low yielding
  Lahontan: resistant to wilt and stem nematode, moderately winter hardy

Birdsfoot trefoil
  Mackinaw, Leo, and Empire: winter hardy, late maturing
  Carroll: winter hardy, medium maturing
  Maitland and Viking: not winter hardy, early maturing

Ladino clover
  Merit, Pilgrim, Regal, Tillman

Red clover
  Lakeland, Arlington, Ottawa, and Florex: winter hardy
  Arlington and Lakeland: anthracnose and mildew resistant

Sweetclover
- *white sweetclover*
  - annual (1-year): Hubam, Israel, Floranna, Emerald
  - biennial (2-year): Denta, Grundy County, Spanish, Willamette, Evergreen, Sangamon, Arctic, Common, (Alpha and Brandon Dwarf are shorter varieties)
- *yellow sweetclover* (all varieties are biennial [2-year]): Madrid, Erector, Aura, Goldtop, Yukon

Kentucky bluegrass
  Merion, Park, Delta, Newport, Common

Smooth bromegrass
  Sac, Baylor and Blair, Fox, Beacon, B-8, Barton, Rebound, Southland

Orchardgrass
  Potomac and Sterling, Dayton and Napier, Hallmark, Able, Comet, Crown, Dart, Hawk, Nordstern, Orbit

Reed canarygrass
  Rise, Frontier, Vantage, Flare

Tall fescue
  Kentucky 31, Kenwell, Kenky, Mo-96

Timothy
  Climax, Lorain, Verdant, Clair, Toro, Timfor, Mortim, Champ

REFERENCES

Anonymous, undated. *Alfalfa*. A guide to production and integrated pest management in the midwest. North Central Regional Extension Publication 113. 224 p.

Ahlgren, G.H. 1956. *Forage Crops*, 2nd ed. McGraw-Hill Book Co., Inc., NY. 536 p.

Chessmore, R.A. 1979. *Profitable Pasture Management*. The Interstate Printers & Publ., Inc., Danville, IL. 424 p.

Donahue, R.L., E.F. Evans, & L.I. Jones. 1956. *The Range and Pasture Book*. Prentice-Hall, Inc., Englewood Cliffs, NJ. 406 p.

Hanson, C.H. (ed.). 1972. *Alfalfa Science and Technology*. American Society of Agronomy, Madison, WI. 812 p.

Heath, M.E., D.S. Metcalfe, & R.F. Barnes. 1973. *Forages*, 3rd ed. Iowa State Univ. Press, Ames, IA. 755 p.

Mays, D.A. (ed.). 1974. *Forage Fertilization*. American Society of Agronomy, Madison, WI. 621 p.

Smith, D. 1962. *Forage Management in the North*. Kendall/Hunt Publ. Co., Dubuque, IA. 219 p.

Willis, H. 1983. *The Rest of the Story. . . About Agriculture Today*. Publ. by the author, Box 692, Wisconsin Dells, WI. 221 p.

INDEX

acidity—see pH

aerobic—10

alfalfa—1, 4-5, 7, 11-13, 19-21, 23, 26-27, 29-30, 33-34, 43-44

alkalinity—see pH

amino acids—16, 18, 26

ammonium—30-31

anaerobic—8, 35-36

animals—5, 7, 15-16, 18-19, 21-24, 27, 29

birdsfoot trefoil—1, 3, 12-13, 42

bluegrass—see Kentucky bluegrass

bromegrass—see smooth bromegrass

calcium—9-11, 18, 20-26, 30, 34, 36-38

canarygrass—see reed canarygrass

chloride—26

companion crops—11-13, 35

deficiency (nutrient)—10, 21-26, 34-38

disease—8, 12, 15, 23, 25, 27, 33-36, 38, 43-44

drought tolerance—3-4, 8, 23

drying (of hay)—30

fertilizers—9-10, 15, 21-24, 26-27, 35-36

fescue—see tall fescue

forages—1, 5, 43-44
    (see also the various kinds)

fungicide-12, 34

genetic engineering—40

glyphosate—40

GM—40-41

grasses—11, 4, 11-12, 30

hardpan—7-9

harvesting—11, 13, 15, 29-31

herbicide—12, 15, 35

humus—8-9, 12-13, 20, 27, 33

insecticide—15, 34

insects—15, 21, 33

Kentucky bluegrass—1, 4, 12-13, 35, 44

Ladino clover—1, 4, 12-13, 44

legumes—1-2, 9, 11-13, 20-21, 30-31, 37

lime—9-11, 21, 26, 30

livestock—see animals

magnesium — 10, 20-21, 23, 26, 34, 36-37

manure—8-10, 12, 25

Monsanto—40-41

nitrate—9, 16, 23, 29-30, 36

nitrogen—2, 10-11, 16, 20, 24, 30-31, 34, 36-38

nitrogen-fixing bacteria —2, 9, 11-12, 20-21, 26, 38
    inoculation—2, 12

nodules—2, 9, 12, 20

orchardgrass—1, 4, 12-13, 42

organic matter—8-10, 36

pesticide—34

pests—8, 12, 21, 25, 27, 33-34, 38, 43-44

pH—8, 10-11, 21-22, 26, 36, 38

phosphate—9-10, 23, 30, 35

phosphorus—9-10, 12, 18, 20, 22-27, 30, 35-38

plowing—11

pollen transfer—40

potash—24-26

potassium—9-10, 12, 20-21, 23-26, 29-30, 34-38

protein—2, 11, 16-17, 19, 21-22, 24, 26, 29

quality—5, 7, 15-27, 29-31, 33, 35, 38

reed canarygrass—i, 4, 13, 44

red clover—1, 3, 12-13, 44

refractometer—18-19, 29-30, 38

roots—2, 4, 7, 9, 21, 23, 30, 35-36

Roundup Ready—40-41

salts—8

seed—12-13, 31

seedbed—11
seeding—11-13, 26, 31
smooth bromegrass—1, 4, 12-13, 44
soil
    aeration—7-8, 12, 20, 27, 33, 35, 38
    conditioners—9
    drainage—7
    erosion—7-8, 11-12
    fertility—2, 7, 11-13, 15, 18, 20, 22,
        27, 30, 33-36
    hardpan—7-9, 35
    humus—8-9, 12-13, 20, 27, 33
    microorganisms—8-9, 12, 20-22, 27,
        33-34, 38
    moisture—12-13, 36
    nutrients—8-9, 11, 18, 20-27, 35-36,
        38
    organic matter—8-9, 36
    pH—8, 10-11, 21-22, 26, 36, 38
    structure—8
    tests—9-10, 21, 25-26, 38
    tilth—8, 35
stand establishment—7-13
sugar—18-19, 29, 35, 38
sulfate—21, 26, 36
sulfur—20-21, 26, 36-37

surfactant—12, 34
sweetclover—1, 4, 13, 44
tall fescue—4, 13, 44
timothy—1, 4, 12-13, 44
tissue analysis—38
toxic
    chemicals—2, 8-9, 12, 20, 33-35
    nutrients—9, 23, 35-36, 38
    soil substances—9-10, 21, 36
trace elements—9, 26-27, 39
    boron—20, 26, 36-37
    cobalt—9, 20, 26
    copper—9, 20, 26, 36-37
    iron—9, 20, 23, 26, 36-37
    manganese—20, 26, 31, 36-37
    molybdenum—9, 20, 26, 37
    zinc—20, 26, 36-37
trefoil—see birdsfoot trefoil
variety—2, 12, 15, 33, 42-43
weather, climate—31, 33, 35-36, 38,
    42-43
weeds—11-12, 15, 25, 35, 38
white clover—1, 4, 12
winter hardiness—3-4, 12, 35, 43-44
yield—12, 15-16, 20, 22, 27, 29-30, 33,
    38, 43-44

Acres U.S.A. — books are just the beginning!

Farmers and gardeners around the world are learning to grow bountiful crops profitably—without risking their own health and destroying the fertility of the soil. *Acres U.S.A.* can show you how. If you want to be on the cutting edge of organic and sustainable growing technologies, techniques, markets, news, analysis and trends, look to *Acres U.S.A.* For more than 35 years, we've been the independent voice for eco-agriculture. Each monthly issue is packed with practical, hands-on information you can put to work on your farm, bringing solutions to your most pressing problems. Get the advice consultants charge thousands for . . .

- Fertility management
- Non-chemical weed & insect control
- Specialty crops & marketing
- Grazing, composting, natural veterinary care
- Soil's link to human & animal health

For a free sample copy or to subscribe, visit us online at

www.acresusa.com

or call toll-free in the U.S. and Canada

1-800-355-5313

Outside U.S. & Canada call 1 (970) 392-4464
fax 1 (970) 392-4464 • info@acresusa.com